International Conference on Mechanical Engineering and Materials Technology (ICMEMT 2015)

Istanbul, Turkey
22-23 August 2015

Editor:

Francisco Marquez

ISBN: 978-1-5108-3260-2

Printed from e-media with permission by:

Curran Associates, Inc.
57 Morehouse Lane
Red Hook, NY 12571

Some format issues inherent in the e-media version may also appear in this print version.

Printed by Curran Associates, Inc. (2016)

For permission requests, please contact Imperial Consortium of Research and Development (ICRD) at the address below.

Imperial Consortium of Research and Development (ICRD)
P.O. Box: 116974
Dubai, United Arab Emirates

Phone: +971 45574074

info@imperialconference.com

Additional copies of this publication are available from:

Curran Associates, Inc.
57 Morehouse Lane
Red Hook, NY 12571 USA
Phone: 845-758-0400
Fax: 845-758-2633
Email: curran@proceedings.com
Web: www.proceedings.com

TABLE OF CONTENTS

Thin-wall Injection Molding Process Optimization Based on CAE Technology

WANG Weidong, SONG Jishun
School of Materials Science Engineering, Tianjin University of Technology, Tianjin 300384, China

Abstract: Plastic products toward "light, thin, short, small" in the direction of the development and thin-wall injection molding technology has gradually become the key technology of the factory. Compared with conventional injection molding, thin-wall injection molding have higher request for injection molding machine and mold. Due to the difficulty of forming, the problem such as breach, flash , weld line, Warping deformation is inevitable. In this paper, the orthogonal test is used to get the best combination of process parameters, and then guide the practical production.

Key words: thin- wall injection molding; warping deformation; orthogonal test

1. Introduction

With the development of science and technology, "light, thin, short, small" plastic parts become "new favorite" of the market, thin -wall molding technology become a new hotspot of research. Thin-wall plastic parts have the advantages of reducing weight, saving raw materials, shortening production cycle, reducing cost, etc. But its theory system has not been formed, but also no special software of numerical simulation of thin-wall injection molding, many related problems in theory and practice is still to be solved. Therefore, this research is of far-reaching significance and function.

2. Definition of thin- wall injection molding

Mahishi and Maloney[1] defined it as: Flow length thickness ratio (L/T) is more than 100 or 150 ;Whetten and Fasset [1] defined it as: the thickness of the plastic parts is less than 1 mm and the projection area of more than 50 cm2; Some scholars define it as: the thickness of the plastic parts is less than 1 mm [2] (or 1.5 mm[3]) or t/d (t-plastic thickness , d-the diameter of the plastic parts , for disc type plastic parts) under 0.05 [4]. Thus it can be seen that of thin-wall injection molding have not unified definition. It is a relative concept, along with the development of technology, the critical value of thin wall injection molding will also be changed. Therefore, the definition of thin wall injection molding should take into account both the thickness of the plastic part, flow length thickness ratio, the projection area, the processing technology and so on.

3. Research object

The plastic part size is 170mm * 110mm * 54mm, and the wall thickness is 0.6mm. Material is polypropylene (PP). Appearance is smooth and beautiful, flawless, dents and other defects, the edge can't have large deformation, forming requirements of plastic parts with certain mechanical properties. From the factory production conditions, the size of the plastic parts and productivity into consideration, hot runner technology can be used, in the form of one module and two cavities.

Using fusion model : triangular unit number is divided into 48770, the node number is 24387, matching rate is 95.1% (> 85%). According to the MPI optimal gate, the gate location is located at the center of the bottom, the melt at the same time reach the ends of the plastic parts.

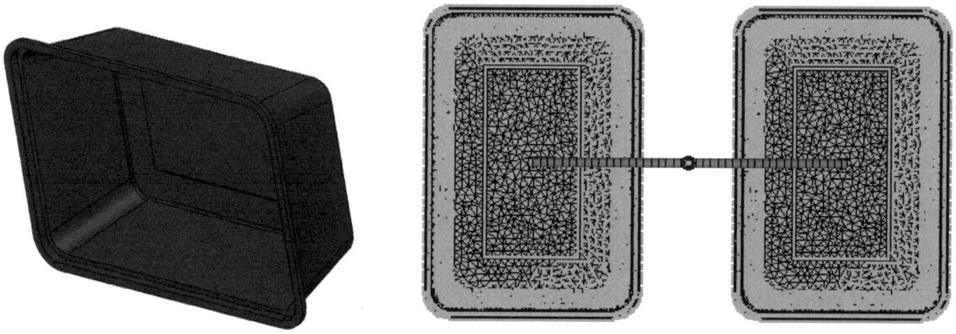

Fig. 1 product graph and simulation model

4. Thin wall injection molding mold design

Mold design is an essential part of in injection molding, the design of reasonable degree directly affect the quality of the product, Thin wall molding products, have a higher request for the mold. In the overall structure of the mold, because it is to bear a higher pressure, the rigidity and the strength of the template is higher. Mould should be designed in tube (draft Angle is 10 ° or so), so as to ensure the accuracy of mold positioning and prevent lateral. As shown in Figure 2, the thickness of the thin wall mold is generally thicker, and the thickness of the moving mold plate and the fixed mold plate is 50mm, the thickness of the convex and concave templates are 100mm. In order to ensure the positioning precision of mould, the convex and concave dies are designed to be in the form of "intensive". The concave part of the concave die is "bite" at the periphery of the concave die. Combining site has 10 ° incline. The mold die fault in under the high injection pressure could be prevented.

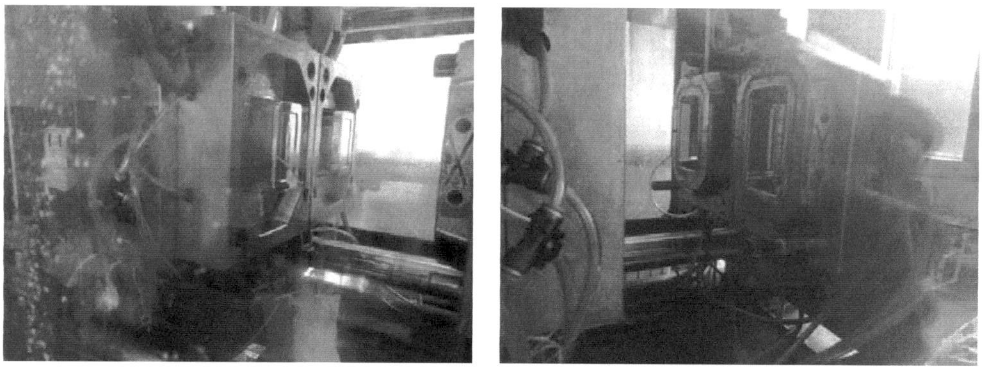

Fig. 2 Mould shape

In casting, runner size should be increased to reduce the pressure drop. But, the volume of condensed material is increased, resulting in waste. Hot runner technology applied to thin-walled injection molding can solve the problem well. The temperature at the main road is set at 260℃, and the temperature at the gate is set to 180-220℃.In terms of cooling, thin-walled plastic parts are more susceptible to warping caused by uneven but cold deformation, affecting dimensional accuracy.

Moreover, the large cooling rate is easy to cause premature solidification of the melt front, resulting in the defect of the injection. In the exhaust side, short filling time, high injection speed, poor exhaust, cause the melt front ignition or short shot defect. In terms of the release, in order to avoid bursting and stick mold, the jack number should be more and bigger sizes.

5. Injection molding CAE analysis

(1) Orthogonal test

Using Moldflow MPI software helps to simulate the process of plastic injection molding, warping deformation as the optimization goal. Experimental use orthogonal experiment which can greatly simplify the number, get the best processing parameters to provide guidance for the actual production. The basic steps of orthogonal experiment: 1) determining the experimental index 2) determining factor and level; 3) using orthogonal table; 4) the experiment and the result analysis.

The main technological parameters that affect the warping deformation include mold temperature, melt temperature, injection time, packing pressure, packing time, cooling time and so on which have a greater impact on the plastic parts of the warping deformation and shrinkage deformation. Experimental factors and levels have shown in table 1.

Table 1 factor level table

Horizontal	mold temperature/℃	melt temperature/℃	packing pressure/Mpa	injection time/s	Packing/s
factor	A	B	C	D	E
1	50	200	50	0.33	0.01
2	60	220	58	0.4	0.02
3	70	240	66	0.47	0.03
4	80	260	74	0.54	0.04

The main indicator of this experiment is to investigate the warpage of plastic parts, its value as small as possible. From the table 2, the sixteenth group of experimental warping is the smallest, as the value is 1.645mm.

Table 2 orthogonal test table

Experiment scheme						Experimental results	
number	A	B	C	D	E	Warping quantity/mm	Volume shrinkage/%
1	1	1	1	1	1	1.868	13.35
2	1	2	2	2	2	1.832	13.57
3	1	3	3	3	3	1.784	14.79
4	1	4	4	4	4	1.685	15.25
5	2	1	2	3	4	1.809	13.29
6	2	2	1	4	3	1.857	13.13
7	2	3	4	1	2	1.696	15.79
8	2	4	3	2	1	1.716	16.32
9	3	1	3	4	2	1.760	13.09

10	3	2	4	3	1	1.745	13.10
11	3	3	1	2	4	1.701	15.25
12	3	4	2	1	3	1.647	16.54
13	4	1	4	2	3	1.902	12.80
14	4	2	3	1	4	1.806	15.11
15	4	3	2	4	1	1.920	14.34
16	4	4	1	3	2	1.645	15.52

Note: A, C, B, D, E, respectively, the mold temperature, melt temperature, packing pressure, injection time, packing time

(2) Range analysis

Range can reflect the effect of various factors, the larger the range, the greater the impact of the factors on the index. From Figure 3, it could be seen that the influence of the plastic part is from the main to the second in this experiment: BAEDC. It is explained that the melt temperature plays a major role in the warping deformation, and the second is the mold temperature. The influence factor of volume shrinkage of plastic parts from the main to second is BCADE, which shows that the melt temperature plays a major role in the volume shrinkage. The optimized process parameters of warping deformation are A3B4C4D3E2, and the process parameters of volume shrinkage are A1B1C4D4E1.

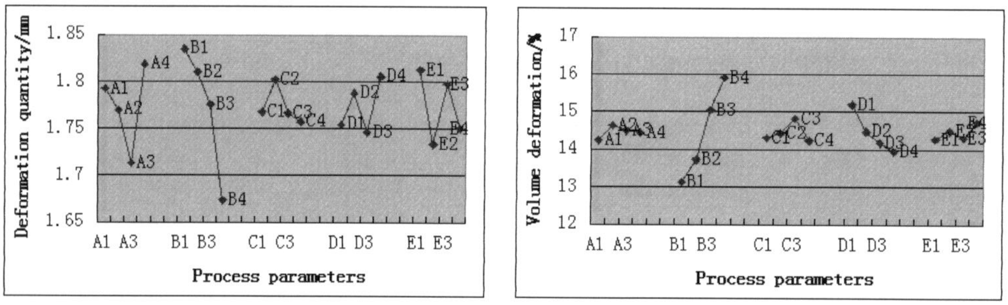

Fig. 4 range line chart

(3) CAE simulation with optimized parameters

According to the process parameters optimization of warping deformation, the combination of A3B4C4D3E2 has simulated again. The simulation results show that the volume shrinkage rate is 18.64%, the warping amount is 1.419mm, and the minimum amount of warping deformation obtained by the orthogonal test is reduced by 13.74%, and the optimal combination is determined, as shown in Figure 5. By adjusting the process parameters, the warping deformation of the plastic part and the product quality is improved.

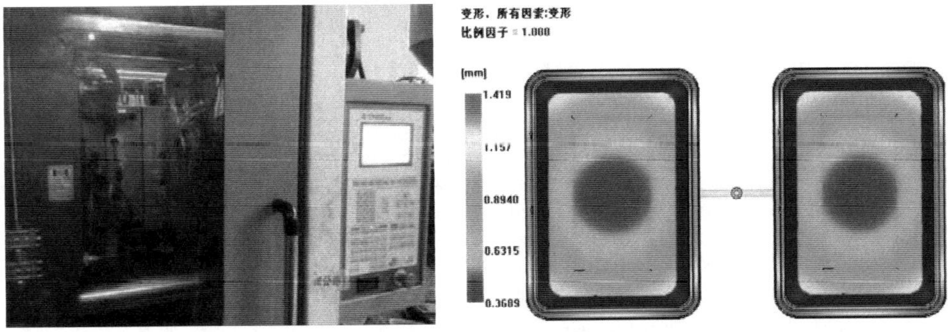

Figure 5 the optimal combination of warping deformation

6. Conclusions

(1) Thin-walled plastic mold needs higher pressure than conventional mold .When the mold design, the use of "concave and convex" form of occlusion, while protecting the mold with a cone.

(2) Moldflow software is used to simulate, and the optimum process parameters are obtained by orthogonal test and range method. When the mold temperature of 70 °C, melt temperature of 260 °C, packing pressure of 74MPa, injection time of 0.47s, packing time of 0.02s, the minimum amount of warpage is 1.623mm.

(3) In the adjustment of injection molding machine parameters, the injection time should be gradually increased to 0.47s, in order to prevent the occurrence of the phenomenon of expansion. Generally not easily adjust the material temperature, because a long time to stabilize after adjusting material temperature, is not conducive to observe and adjust.

Reference

[1] XuGJ.Study of Thin Wall Injection Molding：(Doctor Dissertation)[M].Ohio：The Ohio State University,2004.

[2] Huang Ming-chih,Tai Ching-chin.The Effective Factorsin the Warpage Problem of an Injection-Molded Part with a Thin Shell Feature [J].Materials Processing Technology,2001, 110(1):1-9.

[3] Liao SJ,Chang DY, Chen HJ,etal.Optimal Process Conditions of Shrin kage and Warpage of Thin-Wall Parts[J].Polymer Engineering and Science,2001,44(5):917-928.

[4] 黄虹.塑料成型加工与模具[M].北京:化学工业出版社,2003.

[5] KurtaranH, Ozeelik B, Erzurumlu T. Efficient warpage optimization of thin shell plastic parts using response surface methodology and genetic algorithm [J]. The International Journal of Advanced Manufacturing Technology, 2006, 27(5): 468 -472.

[6] 曾亚森,何和智.基于翘曲的注塑成型工艺参数优化组合[J].高分子材料科学与工程，2009, 25(6): 163 -166.

*** Formatting Issue - Best Available Paper/Graphic ***

Advanced in the Experimental Methods of Tensile Mechanical Test of Single Plant Fiber

Wentao Chen[1, a], Bing Fang[1, b], Dai Qunhua[1, c], Dapeng Ye[1, d]

[1]College of Mechanical and Electrical Engineering, Fujian Agriculture and Forestry University, Fuzhou, Fujian, China

[a]winterchencool@163.com, [b]fangbing008@163.com,[c]898125225@qq.com, [d]fau_ydp@163.com

Abstract: Mechanical properties of single plant fiber provide a more precise theoretical explanation for their macroscopic mechanical behavior analysis. This paper described the development of tensile mechanical properties of single plant fiber in domestic and foreign research, reviewed the process of single fiber tensile test in detail, and discussed some basic principles of study about the mechanical properties of single plant fiber tensile. In order to provide information and reference for future research, the existing insufficient and outlook on the mechanical properties of single plant fiber tensile have also been proposed.

Keywords: Single fiber, tensile technique, mechanical properties, nanoindentation technique;

1. Introduction

Plant fiber is a kind of sclerenchyma which is widely present in plants. Its cell structure is slender, jagged ends of the cell, with a thick secondary wall. In plants, fibers mainly play a mechanical support role. The studies on mechanical characteristics of single plant fiber provide a theoretical basis for large agricultural cutting machinery on design and manufacture. Meanwhile, the plant fiber is the most important component unit of composite green materials, giving the composite product good mechanical properties. Mechanical properties analysis of single plant fibers has been gradually developed in recent years, which play an important part on Plant Biomechanics. There is a huge leap between the macro mechanical analysis and the micro mechanics analysis on the cellular level. Because the plant fiber shape is very small, its demanding conditions for the mechanical analysis techniques and measurement equipment and so on are much higher. Currently, the tensile mechanical technology of single plant fiber involved single fiber drawing technology, nano-indentation technique and zero-span stretching techniques. From tensile mechanical properties of single plant fiber and theoretical knowledge point of view, this paper expounded research results and test methods at home and abroad by using of these three technologies and research, then put forward a number of recommendations for tensile mechanical properties of single plant fiber on future research.

2. The Mechanical Theory of Single Fiber Tensile

At the beginning of the tensile test, stretching jig at the set parameters applied the tension force to the single plant fiber. Once the fiber stretched destructively the loading stopped immediately, and a corresponding load-displacement curve would be outputted. Under the stretching force F on length L, the stress of a single fiber with a cross section area S is F / S, represented by σ. The ratio of displacement variable quantity L and the initial length L is called strain, represented by ε. The ratio of stress and strain calls elastic modulus, represented by E. Tensile modulus, also known as Young's modulus, is one of the most common elastic modulus [1]. According to the material mechanics knowledge, the basic formula is:

$$E = \sigma/\varepsilon = F\quad L/SL \qquad (1\text{-}1)$$

Where E is Young's modulus in tensile (MPa). L is the displacement variable quantity (μm), L is the initial length of single fiber (μm), S is the cross section area of single fiber (mm^2), and F is the stretching force (mN).

3. Single Fiber Tensile Technology

Single fiber tensile technology is a kind of direct axial stretching techniques. In order to attain some important indicators of the cell wall, such as the longitudinal elastic modulus, tensile strength, hardness *etc.*, single fibers should be isolated by chemical or mechanical methods. Jayne is one of the earliest single fiber tensile technology research scholars [3]. However, due to the insufficient of the test equipment conditions, Jayne's study hadn't cause concern. 1970s, Page published reports on the mechanical properties of single softwood tracheid and its determination method, which had aroused great interest in the field of pulp and paper and wood science[4]. In the single fiber tensile techniques, fibers preparing, clamping, orientation, cell wall cross-sectional area and the filaments angle measurement are the difficulties in the test [2].

4. Preparation of Single Fibers

Preparation of conventional single fibers is aimed to oxidize the lignin by the chemical drugs for isolating fiber bundle. On chemical methods, hydrogen peroxide or chlorite and acetic acid with a certain percentage are generally used to get the segregation liquid. After a long soak, single fibers were softened and isolated. Elsa used such method to isolate single pulp fiber, and took single fiber tensile test in the air and water respectively [5]. Zhang took the single fir wood fiber for the study, using three kinds of chemical methods to remove lignin in different degrees at the cellular level, and investigated the effect of lignin on fiber tensile mechanical properties of a timber single [6]. With the high precision of mechanical electronic measuring instruments appearing, Burgert proposed the technology using mechanical stripping means firstly to prepare single plant fiber, and through two different separating methods to test the changes on microfibril angle of wood fibers, which resulted that the mechanical stripping means could avoid causing a change in cell wall structure and chemical properties of the fibers compared with the traditional chemical isolation method [7].

Compared to the advantages, simple, fast preparation and complete separation, of the chemical preparation method, the mechanical stripping means required expensive equipment, complicated operation, which was not easy to grasp. In actual production or research, it should be combined with chemical methods and physical methods to achieve the purpose of the preparation of the individual fibers, and the results would be best.

5. The Clamping Methods of Single Fiber

There are three main methods to grip a single fiber. They are mechanical clamping method, viscose method and ball groove clamping methods. Jayne used mechanical clamping method to separate the wood fiber on the tensile mechanical properties test. The results indicated the sources of variation was not just the fiber itself size and nature of heterogeneity, but also may be caused by the process of drawing method and improper sampling operations [8]. Mechanical clamping method is easy to cause stress concentration of fiber in the holder so the probability of fiber producing shear fracture increases, which would result in greater error in the test results.

Viscose method is first fixed fiber with a certain adhesive on the easy-loading medium, and then placed in a tensile fixture to stretch. According to different medias, paper, cellulose acetate, plastic or metal sheet *etc.* are adopted [9]. While the adhesive is commonly used by two-component epoxy resin. Different adhesive can also cause the variable of fibers mechanical properties. Therefore, how to choose a suitable adhesive is very important, which would ensure the adhesive does not penetrate into the fiber and maintain the stability well between the fiber and media so that fibers were not slip or withdrawn.

*** Formatting Issue - Best Available Paper/Graphic ***

Kersavage proposed groove ball gripping method in the 1970s which was on the basis of the adhesive [10]. The method of operation first dropped epoxy glue on both ends of the fiber drops, and then slotted it into the ball fixture system to tensile test. Two small ball-like gum drops can be cured at free rotation in the slot of the ball groove. This method not only solves the problem of fiber on crushing and slips, but also solves the problem of free orientation of a single fiber, reducing the damage caused by the tensile shear. Huang *et al* (2009) used the ball groove clamping method to solve difficulties on plant staple holder, and fiber twisting and shear problems during testing [11].

6. The Test Equipment of Mechanical Properties of Single Fiber

Because the volume of individual fibers is very small, high-precision electronic measuring equipment are required to observe cell morphology of single fiber to get the length, the cross-sectional area, microfibril angle and other mechanical parameters of single fiber.

Cross-sectional area is an important parameter to calculate strain, Goom *et al* (2002) have made a great contribution in wall area measurements of single fiber by using a laser scanning confocal microscope to get the cell wall area of single fiber directly, which had improved speed and accuracy on the measurement [9].

According to the principle of wetting force measurement, Chen *et al* (2011) used contact angle measuring instrument Kruss K100MK2 to measure contact angles of several single plant fibers [12]. It is a simple that can be used to determine the most bamboo and other natural plant contact angle of single fiber.

Wang *et al* (2011) used self-developed high precision fiber mechanical properties Tester (SF-I) to study the longitudinal tensile properties and water dependence of single fiber of bamboo, fir, ramie [13]. The device has a dedicated fixture for plant short fiber and can quickly and easily measure the longitudinal length of the mechanical properties of various plant short fibers whose length is in more than 1.2mm. This kinds of device are also equipped with micro-climate control system for studying the influence of moisture content of the plant fiber on longitudinal tensile properties. Liu *et al* (2012) developed a high precision experimental apparatus of tensile mechanical properties of micro-fiber material named FMT-I, which were used for micro-scale mechanical properties test and scale effect of experimental research, and it is based on LabVIEW software platform to develop the corresponding data acquisition and control system, achieving fully automated on testing process [14].

7. Nanoindentation Technique

Nano indentation technique is also known as depth-sensitive indentation technique, which have a specific shape of small diamond indenter (tip radius of curvature is typically less than 100nm) pressing into the material surface, and continuously measures the loading and unloading process acting on the indentation depth of sample and the pressure of the needle. Load in this mode is typically only a few micro-Newton. The maximum indentation depth is only a few microns, which are an ideal method to detect mechanical properties of small volumes. Cheng *et al* think Nano indentation system need to develop more sophisticated and can reduce or eliminate the stress of the probe, exploring the more simple and convenient sample preparation techniques so that eliminate the influence of changes in the indentation area [15] .

The fundamental difference between the Nano indentation technology and traditional hardness testing techniques is that Nano indentation technique is obtained by the empirical formula, not based on the photos of the contact area after unloading indentation. Nano indentation can finish a variety of mechanical properties test, such as hardness, modulus, yield stress, fracture toughness, creep and so on, particularly in hardness and elastic modulus measurements, the measurement error is small, with high spatial resolution. It's adapted to detect micron or submicron scale of thin films or micro material on the test of mechanical properties [16].However, nanotechnology can only get the material modulus and hardness, the strength information cannot be obtained. As

*** Formatting Issue - Best Available Paper/Graphic ***

for the apparent creep and relaxation phenomena appeared on viscoelastic material and non-homogeneous anisotropic material, its measurement repeatability and accuracy remains to be further improved.

8. Zero-span Tensile Testing Technique

Zero-span tensile testing technique refers to the paper's entirely fault by the interface across on both ends of the fixture due to fiber breakage at the condition of zero state stretching, excluding the influence of tensile fracture when fiber is pulled on normal pitch, which can be used directly evaluate the tensile strength of the fibers.Yu *et al* (2006) took the areas of the paper-making technology into the field of zero-span tensile wood science for the rapid evaluation of longitudinal tensile strength of tracheid of the plantation Chinese firm[17]. The results show that the use of zero-span tension technology has the characteristics of rapid measurement, strong operability *etc*, which can be used as a new method to quickly evaluate the longitudinal mechanical properties of wood fibers. Shangguan *et al* (2011) commented zero-span tension technology respectively from slip and clamping force, the specimen thickness and moisture content of the sample and reckoned that researchers should determine the appropriate clamping force and thickness of the sample and so on depending on the material to finally get the accurate measurement results[18].

Zero-span tensile testing technique has the advantages of the simple preparation of sample, fast measurement speed, good repeatability and so on. However, the problem of stress concentration of the clamping also restricts the zero-span tensile, and the zero-span tensile can only get the tensile strength, which provides the relatively simple mechanical parameters.

9. Prospect and Conclusion

How to separate a single fiber without destroying the chemical composition of fiber cell wall itself will be a difficult problem in the directions of research in the future. The single fiber tensile technology could be combined with other technologies, such as the combination with high-precision instruments to improve the efficiency and accuracy of the test has important research prospects.

The analysis of mechanical properties of single fiber tensile is at an early stage. There is no in-depth analysis of the change in the chemical composition of the cell wall and the effects of the underlying mechanisms of crack of fibers. When the external environment (such as temperature, humidity, PH value) changes, it will affect the mechanical properties of single fibers, thus there is a urgent need to study factors affecting in this area.

Plant wood of plant cells during the process of the plant cell development is an important change on the process of cell wall material. How to regulate and modify in the growth or processing of the plant will be an important direction of research on getting ideal composite fiber material, which is also of great value for research about genetics and breeding and environmental protection *etc*.

References

[1] Cai Lili,Wang Huibin.The Improvement of Young's Modulus Experiment Device for Measuring Metals by Stretch Method[J].Physical Experiment of College,2015,01:57-59.

[2] Wang Yonghong,Wang Rui,Gao Longya et al.Mechanical Properties of Single Daemonorops margaritae Fiber[J].Journal of Northeast Forestry University,2012,12:87-89+107.

[3] Jayne B A. Some mechanical properties of wood fibers in tension[J]. Forest products journal.1960:316-322

[4] Page D H, El-Hosseiny F, Winkler K. Behaviour of single wood fibers under axial tensile strain[J],Nature, 1971, 229(5282): 252-253.

[5] Elsa L,Ehmrooth,Petter K.The tensile testing of single wood pulp fibers in air and in water[J].Wood and Fiber Science,1984,16(4)： 549-566.

[6] Zhang Shuangyan,Fei Benhua,Yu Yan.Influence of lignin content on tensile properties of single wood

*** Formatting Issue - Best Available Paper/Graphic ***

fiber[J].Journal of Beijing Forestry University,2012,34:131-134.

[7] Burgert I,Keckes J,Fruhmann K,et a1.A comparison of two techniques for wood fiber isolation evaluation by tensile tests on single fibers with different microfiber angle[J].Plant Biol.4:9-12.

[8] Jayne B A. Some mechanical properties of wood fibers in tension[J]. Forest products journal. 1960:316-322.

[9] Cao Shuangping.The measurement and evaluation of tensile properties of plant individual fibers[D].Chinese Academy of Forestry,2014.

[10] Kersavage P C. A systemfor automatically recording the load-elongation characteristics of single wood fibers under controlled relative humidity conditions[J]. USDA. U. S. Government Printing Office, 1973

[11] Huang Yanhui,Fei Benhua,Yu Yan et al.Study on the Mechanical Properties of Moso Bamboo Single Fiber[J]. China pulp & paper,2009,08:10-12.

[12] Chen Xiaomeng,Wang Ge,Cheng Haitao.Study on force method for measurement of single fiber contact angle[J].Journal of central south university of forestry & technology,2011,04:192-195.

[13] Wang Ge,Yu Yan,Cao Shuangping et al.Longitudinal Tensile Properties Determination of Vegetable Single Fibers by the Micro-Tensile Technology[J].Scientia Silvae Sinicae,2011,07：151-155.

[14] Liu Dabiao,He Yuming,Hu Peng et al.On the Mechanical Properties in Micro-Tensile Testing of Single Fibe[J]. Journal of Experimental Mechanics,2012,01:61-69.

[15] Cheng Xianbao,Wang Xiaoqing,Yu Yan et al.Application of Nanoindentation Technology in Determining the Mechanical Properties of Wood Cell Wall[J].World Forestry Research,2011,05：40-46.

[16] Niu Eryan,Zhang Longfei,Chen Zhilin et al.The nano-indentation technique in testing the mechanical properties of modified wood[J].Wood Processing Machinery,2015,02:52-56+38.

[17] Yu Yan,Fei Benhua,Zhang Bo.Estimation of Longitudinal Tensile Strength of Tracheids with Zero-Span Tension Technique[J].Scientia Silvae Sinicae,2006,07:83-86.

[18] ShangGuan Weiwei,Xing Xinting,Fei Benhua et al.Advances in the Experimental Methods of Mechanical Test of Wood Cell Wall[J].Journal of Northwest Forestry University,2011,06：149-153.

Phantom Track Discrimination for Radar Network Based on Position Deviation

ZHONG YANG,GUOHONG WANG,DIANXING SUN
Institute of Information Fusion Technology Department
Naval Aeronautical and Astronautical University
Yantai,264001
CHINA
harrynil@163.com

Abstract: Due to the difficulty for radar network to discriminating phantom track deception released by jamming equipments, a discrimination method based on position deviation proposed in Earth-Centred, Earth-Fixed (ECEF) coordinate system. Firstly, position error of radar network and its influence on phantom track measurement is explored. We also give the exact statistical distribution of position deviation in ECEF coordinate system, according to which the phantom track is discriminated by hypothesis testing. Moreover, 2/3 logic is carried out to enhance the reliability of the judgment. Lastly, relationships between correct-discriminating probability and bias error or radar measurement error studied through simulation. The results of simulation and practical test verify the correctness of the analysis.

Key-Words: Radar Network, Phantom Track Deception, ECEF, Position Deviation, 2/3 logic, Hypothesis Testing

1 Introduction

As an important aspect of Electronic Warfare (EW), radar counter-counter-measures in modern battlefield are getting more and more considerable attention[1-5]. A technique which produces alignment phantom track with unmanned aerial vehicles (UAVs) to deceive radar network is proposed in Ref.[6]and[7]. This novel way of interference is different from the ordinary distance or angle deceptive jamming, because it can produce stable tracks in radar network fusion center, creating more harm and stronger delusion. In order to implement phantom track jamming, the enemy must first precisely locate radar position. TDOA measurements are used to locate radar position in the polar coordinates in Ref.[8]. On this basis, phantom track jamming to surveillance radar and mono-pulse three-coordinate radar are studied in Ref.[9]and[10]. Phantom track jamming technique is transformed into two-ends constrained optimization problem in Ref.[11], which makes cooperative track deception come true. To create phantom tracks similar to real tracks, parameters of radars such as signal type, antenna rotating speed, width of main lobe, pulse interval, signal amplitude etc should be reconnaissance, and which should be very accurate in the process of signal modulation and transmission. But in fact, there will always be a variety of bias, such as radar location error, radar characteristic measurement error, etc. These biases make it possible to discriminate phantom track jamming with the help of appropriate data processing algorithm. The like-hold testing method is used to discriminate phantom track in Ref.[12], which makes use of random error produced by ECAVs(Electronic Combat Air Vehicles) formation flight control. Hypothesis testing is applied to identify phantom tracks in Ref.[13], which makes use of the inspection of the innovation extracted from target tracking process. Statistical analysis is utilized to testify covariance matrix in Ref.[14]. Fewer papers consider the fact that there are radar position deviations during the locating of radar position. In the presence of this problem, this paper analyses the effect of position deviation to false target measurement, with which basis phantom track could be identified by hypothesis testing.

2 Problem Formulation

As seen from Figure 1, assuming that the radar network consists of N radars, and each radar is deceived by one jamming equipment. When radar performs, jamming equipments capture radar signal, copy, modulate and transmit it back to radar, and produce phantom track by generating deceiving range and deceiving angle in radar polar coordinate system.

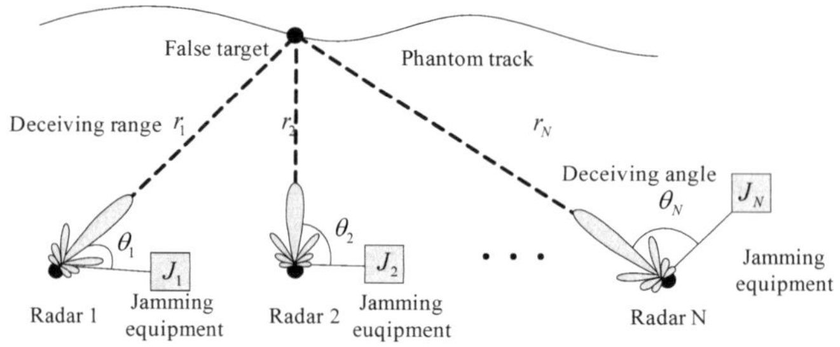

Fig.1 Radar network and phantom track

The target measurement in radar i polar coordinate system at time of k is denoted by $(\rho_i(k),\theta_i(k),\varphi_i(k))$ $(i=1,2,\cdots,\mathrm{N};k=1,2,\cdots,\mathrm{M})$,and is given by:

$$\begin{cases}\rho_i(k)=\rho^i(k)+\eta_\rho^i+\varsigma_\rho^i\\ \theta_i(k)=\theta^i(k)+\eta_\theta^i+\varsigma_\theta^i\quad(i=1,2,\cdots,\mathrm{N};k=1,2,\cdots,\mathrm{M})\\ \varphi_i(k)=\varphi^i(k)+\eta_\varphi^i+\varsigma_\varphi^i\end{cases}\tag{1}$$

where $\left(\rho^i(k),\theta^i(k),\varphi^i(k)\right)$ is the true target position in polar coordinate system, $(\eta_\rho^i,\eta_\theta^i,\eta_\varphi^i)$ is bias error produced by radar position error, $(\varsigma_\rho^i,\varsigma_\theta^i,\varsigma_\varphi^i)$ is measurement error, and $\varsigma_\rho^i\sim N\left(0,\sigma_{\rho i}^2\right)$, $\varsigma_\theta^i\sim N\left(0,\sigma_{\theta i}^2\right)$, and $\varsigma_\varphi^i\sim N\left(0,\sigma_{\varphi i}^2\right)$.

The measurement in radar polar coordinate system can be transformed into radar East-North-Up (ENU) coordinate system by[15]:

$$\begin{cases}X_i(k)=\rho_i(k)\cos\varphi_i(k)\cos\theta_i(k)\\ Y_i(k)=\rho_i(k)\cos\varphi_i(k)\sin\theta_i(k)\quad(i=1,2,\cdots,\mathrm{N};k=1,2,\cdots,\mathrm{M})\\ Z_i(k)=\rho_i(k)\sin\varphi_i(k)\end{cases}\tag{2}$$

where $\left(X_i(k),Y_i(k),Z_i(k)\right)$ is target measurement in radar i 's ENU coordinate system.

The measurement in radar ENU coordinate system can be transformed into ECEF coordinate system by:

$$\begin{bmatrix}X_{ECEF,i}(k)\\ Y_{ECEF,i}(k)\\ Z_{ECEF,i}(k)\end{bmatrix}=\boldsymbol{R}_i\begin{bmatrix}X_i(k)\\ Y_i(k)\\ Z_i(k)\end{bmatrix}+\begin{bmatrix}X_{R,i}\\ Y_{R,i}\\ Z_{R,i}\end{bmatrix}(i=1,2,\cdots,\mathrm{N};k=1,2,\cdots,\mathrm{M})\tag{3}$$

where $\left(X_{ECEF,i}(k),Y_{ECEF,i}(k),Z_{ECEF,i}(k)\right)$ is target measurement in ECEF coordinate system, $\left(X_{R,i}(k),Y_{R,i}(k),Z_{R,i}(k)\right)$ is radar position in ECEF coordinate system.

Assuming that the longitude, latitude and height of radar i is denoted by $(L_{R,i},B_{R,i},H_{R,i})(i=1,2,\cdots,\mathrm{N})$, then one has

$$\begin{cases}X_{R,i}=\left[N_R(1-\mathrm{e}_1^2)+H_{R,i}\right]\sin(B_{R,i})\\ Y_{R,i}=\left(N_R+H_{R,i}\right)\cos(B_{R,i})\cos(L_{R,i})(i=1,2,\cdots,\mathrm{N})\\ Z_{R,i}=\left(N_R+H_{R,i}\right)\cos(B_{R,i})\sin(L_{R,i})\end{cases}\tag{4}$$

where $e_1^2=\dfrac{a^2\text{-}b^2}{a^2}$ is first eccentricity, $N_R=\dfrac{a}{\sqrt{1\text{-}e_1^2\sin^2B_i}}$, a=6378137m,b=6356752m in WGS-84 coordinate system.

The conversion matrix $\boldsymbol{R}_i$ in Eq.(3) is:

$$\boldsymbol{R}_i=\begin{bmatrix}-\sin L_{R,i} & -\sin B_{R,i}\cos L_{R,i} & \cos B_{R,i}\cos L_{R,i}\\ \cos L_{R,i} & -\sin B_{R,i}\sin L_{R,i} & \cos B_{R,i}\sin L_{R,i}\\ 0 & \cos B_{R,i} & \sin B_{R,i}\end{bmatrix}(i=1,2,\cdots,\mathrm{N})\tag{5}$$

where $(L_{R,i}, B_{R,i}, H_{R,i})$ is radar i 's geographical coordinate.

3 Model Analysis

When two different radars in the same radar network are tracking one same target, because of the existence of radar position error, there are bias errors in radar measurements in ECEF coordinate system. Assuming that bias error in ECEF coordinate system is denoted by:

$$\begin{cases} \Delta X(i,j) = X_{\text{ECEF},i} - X_{\text{ECEF},j} \\ \Delta Y(i,j) = Y_{\text{ECEF},i} - Y_{\text{ECEF},j} \\ \Delta Z(i,j) = Z_{\text{ECEF},i} - Z_{\text{ECEF},j} \end{cases} \tag{6}$$

Substituting Eq.(3) into (6) yields:

$$\begin{cases} \Delta X(i,j) = X_{R,i} - X_{R,j} + r_{11}X_i + r_{12}Y_i + r_{13}Z_i - (r_{11}^{'}X_j + r_{12}^{'}Y_j + r_{13}^{'}Z_j) \\ \Delta Y(i,j) = Y_{R,i} - Y_{R,j} + r_{21}X_i + r_{22}Y_i + r_{23}Z_i - (r_{21}^{'}X_j + r_{22}^{'}Y_j + r_{23}^{'}Z_j) \\ \Delta Z(i,j) = Y_{R,i} - Y_{R,j} + r_{31}X_i + r_{32}Y_i + r_{33}Z_i - (r_{31}^{'}X_j + r_{32}^{'}Y_j + r_{33}^{'}Z_j) \end{cases} \tag{7}$$

where $r_{11}, r_{12}, r_{13}, r_{21}, r_{22}, r_{23}, r_{31}, r_{32}, r_{33}$ are elements in conversion matrix $\boldsymbol{R}_i$, $r_{11}^{'}, r_{12}^{'}, r_{13}^{'}, r_{21}^{'}, r_{22}^{'}, r_{23}^{'}, r_{31}^{'}, r_{32}^{'}, r_{33}^{'}$ are elements in conversion matrix $\boldsymbol{R}_j$.

For simplicity of the expression, let:

$$f_X(\rho,\eta,\varsigma) = r_{11}X_i + r_{12}Y_i + r_{13}Z_i - (r_{11}^{'}X_j + r_{12}^{'}Y_j + r_{13}^{'}Z_j) \tag{8}$$

and

$$f_Y(\rho,\eta,\varsigma) = r_{21}X_i + r_{22}Y_i + r_{23}Z_i - (r_{21}^{'}X_j + r_{22}^{'}Y_j + r_{23}^{'}Z_j) \tag{9}$$

Substituting Eq.(1), (2) into (8) results in:

$$\begin{aligned} f_X(\boldsymbol{\rho},\boldsymbol{\eta},\varsigma) &= r_{11}(\rho^i + \eta_\rho^i + \varsigma_\rho^i)\cos(\varphi^i + \eta_\varphi^i + \varsigma_\varphi^i)\cos(\theta^i + \eta_\theta^i + \varsigma_\theta^i) + r_{12}(\rho^i + \eta_\rho^i + \varsigma_\rho^i)\cos(\varphi^i + \eta_\varphi^i + \varsigma_\varphi^i)\sin(\theta^i + \eta_\theta^i + \varsigma_\theta^i) + \ldots \\ & r_{13}(\rho^i + \eta_\rho^i + \varsigma_\rho^i)\sin(\varphi^i + \eta_\varphi^i + \varsigma_\varphi^i) - \left(r_{11}^{'}(\rho^j + \eta_\rho^j + \varsigma_\rho^j)\cos(\varphi^j + \eta_\varphi^j + \varsigma_\varphi^j)\cos(\theta^j + \eta_\theta^j + \varsigma_\theta^j) + \ldots\right. \\ & \left. r_{12}^{'}(\rho^j + \eta_\rho^j + \varsigma_\rho^j)\cos(\varphi^j + \eta_\varphi^j + \varsigma_\varphi^j)\sin(\theta^j + \eta_\theta^j + \varsigma_\theta^j) + r_{13}^{'}(\rho^j + \eta_\rho^j + \varsigma_\rho^j)\sin(\varphi^j + \eta_\varphi^j + \varsigma_\varphi^j)\right) \end{aligned} \tag{10}$$

where $\boldsymbol{\rho} = [\rho^i, \theta^i, \varphi^i, \rho^j, \theta^j, \varphi^j]^T$, $\boldsymbol{\eta} = [\eta_\rho^i, \eta_\theta^i, \eta_\varphi^i, \eta_\rho^j, \eta_\theta^j, \eta_\varphi^j]^T$, and $\varsigma = [\varsigma_\rho^i, \varsigma_\theta^i, \varsigma_\varphi^i, \varsigma_\rho^j, \varsigma_\theta^j, \varsigma_\varphi^j]^T$.

We can make Taylor expand on $f_X(\boldsymbol{\rho},\boldsymbol{\eta},\varsigma)$ at $\boldsymbol{\rho} = \boldsymbol{0}$ and $\boldsymbol{\eta} = \boldsymbol{0}$:

$$f_X(\boldsymbol{\rho},\boldsymbol{\eta},\varsigma) \approx f_X(\boldsymbol{\rho},\boldsymbol{\eta}',\varsigma') + \frac{\partial f_X(\boldsymbol{\rho},\boldsymbol{\eta}',\varsigma')}{\partial \boldsymbol{\eta}}(\boldsymbol{\eta} - \boldsymbol{\eta}') + \frac{\partial f_X(\boldsymbol{\rho},\boldsymbol{\eta}',\varsigma')}{\partial \varsigma}(\varsigma - \varsigma') \tag{11}$$

where $\boldsymbol{\eta}' = \varsigma' = [0,0,0,0,0,0]^T$, and

$$\begin{aligned} f_X(\boldsymbol{\rho},\boldsymbol{\eta}',\varsigma') &= r_{11}\rho^i\cos\varphi^i\cos\theta^i + r_{12}\rho^i\cos\varphi^i\sin\theta^i + r_{13}\rho^i\sin\varphi^i - \ldots \\ & \left(r_{11}^{'}\rho^j\cos\varphi^j\cos\theta^j + r_{12}^{'}\rho^j\cos\varphi^j\sin\theta^j + r_{13}^{'}\rho^j\sin\varphi^j\right) \end{aligned} \tag{12}$$

$$\begin{aligned} \frac{\partial f_X(\boldsymbol{\rho},\boldsymbol{\eta}',\varsigma')}{\partial \boldsymbol{\eta}}(\boldsymbol{\eta} - \boldsymbol{\eta}') &= \left(r_{11}\cos\varphi^i\cos\theta^i + r_{12}\cos\varphi^i\sin\theta^i + r_{13}\sin\varphi^i\right)\boldsymbol{\eta}_\rho^i + \left(-r_{11}\rho^i\cos\varphi^i\sin\theta^i + r_{12}\rho^i\cos\varphi^i\cos\theta^i\right)\eta_\theta^i + \ldots \\ & \left(-r_{11}\rho^i\sin\varphi^i\cos\theta^i - r_{12}\rho^i\sin\varphi^i\sin\theta^i + r_{13}\rho^i\cos\varphi^i\right)\eta_\varphi^i + \left(r_{11}^{'}\cos\varphi^j\cos\theta^j + r_{12}^{'}\cos\varphi^j\sin\theta^j + r_{13}^{'}\sin\varphi^j\right)\boldsymbol{\eta}_\rho^j + \ldots \\ & \left(-r_{11}^{'}\rho^j\cos\varphi^j\sin\theta^j + r_{12}^{'}\rho^j\cos\varphi^j\cos\theta^j\right)\eta_\theta^j + \left(-r_{11}^{'}\rho^j\sin\varphi^j\cos\theta^j - r_{12}^{'}\rho^j\sin\varphi^j\sin\theta^j + r_{13}^{'}\rho^j\cos\varphi^j\right)\eta_\varphi^j \end{aligned} \tag{13}$$

and

$$\begin{aligned} \frac{\partial f_X(\boldsymbol{\rho},\boldsymbol{\eta}',\varsigma')}{\partial \varsigma}(\varsigma - \varsigma') &= \left(r_{11}\cos\varphi^i\cos\theta^i + r_{12}\cos\varphi^i\sin\theta^i + r_{13}\sin\varphi^i\right)\varsigma_\rho^i + \left(-r_{11}\rho^i\cos\varphi^i\sin\theta^i + r_{12}\rho^i\cos\varphi^i\cos\theta^i\right)\varsigma_\theta^i + \ldots \\ & \left(-r_{11}\rho^i\sin\varphi^i\cos\theta^i - r_{12}\rho^i\sin\varphi^i\sin\theta^i + r_{13}\rho^i\cos\varphi^i\right)\varsigma_\varphi^i + \left(r_{11}^{'}\cos\varphi^j\cos\theta^j + r_{12}^{'}\cos\varphi^j\sin\theta^j + r_{13}^{'}\sin\varphi^j\right)\varsigma_\rho^j + \ldots \\ & \left(-r_{11}^{'}\rho^j\cos\varphi^j\sin\theta^j + r_{12}^{'}\rho^j\cos\varphi^j\cos\theta^j\right)\varsigma_\theta^j + \left(-r_{11}^{'}\rho^j\sin\varphi^j\cos\theta^j - r_{12}^{'}\rho^j\sin\varphi^j\sin\theta^j + r_{13}^{'}\rho^j\cos\varphi^j\right)\varsigma_\varphi^j \end{aligned} \tag{14}$$

Combining Eq.(7) and (10) results in:

$$\Delta X(i,j) \approx X_{R,i} - X_{R,j} + f_X(\boldsymbol{\rho},\boldsymbol{\eta}',\varsigma') + \frac{\partial f_X(\boldsymbol{\rho},\boldsymbol{\eta}',\varsigma')}{\partial \boldsymbol{\eta}}(\boldsymbol{\eta}-\boldsymbol{\eta}') + \frac{\partial f_X(\boldsymbol{\rho},\boldsymbol{\eta}',\varsigma')}{\partial \varsigma}(\varsigma - \varsigma') \tag{15}$$

where $f_X(\boldsymbol{\rho},\boldsymbol{\eta}',\varsigma')$, $\frac{\partial f_X(\boldsymbol{\rho},\boldsymbol{\eta}',\varsigma')}{\partial \boldsymbol{\eta}}(\boldsymbol{\eta}-\boldsymbol{\eta}')$, $\frac{\partial f_X(\boldsymbol{\rho},\boldsymbol{\eta}',\varsigma')}{\partial \varsigma}(\varsigma-\varsigma')$ is written in Eq.(11), (12) and (13).

For real target, there is no bias error, if measurement error is zero, then $\Delta X(i,j)=0$, $X_{R,i} - X_{R,j} + f_X(\boldsymbol{\rho},\boldsymbol{\eta}',\varsigma') = 0$, so for real target:

$$\Delta X(i,j) \approx \frac{\partial f_X(\boldsymbol{\rho},\boldsymbol{\eta}',\varsigma')}{\partial \varsigma}(\varsigma-\varsigma') \tag{16}$$

For false target:

$$\Delta X(i,j) \approx \frac{\partial f_X(\boldsymbol{\rho},\boldsymbol{\eta}',\varsigma')}{\partial \boldsymbol{\eta}}(\boldsymbol{\eta}-\boldsymbol{\eta}') + \frac{\partial f_X(\boldsymbol{\rho},\boldsymbol{\eta}',\varsigma')}{\partial \varsigma}(\varsigma-\varsigma') \tag{17}$$

As can be seen from Eq.(17), false target position error consists of two parts, one is root deviation produced by radar position error, the other is radar measurement error ς. And

$$\begin{aligned} f_X(\boldsymbol{\eta}) = {} & \frac{\partial f_X(\boldsymbol{\rho},\boldsymbol{\eta}',\varsigma')}{\partial \boldsymbol{\eta}}(\boldsymbol{\eta}-\boldsymbol{\eta}') = \left(r_{11}\cos\varphi^i\cos\theta^i + r_{12}\cos\varphi^i\sin\theta^i + r_{13}\sin\varphi^i\right)\eta_\rho^i + \ldots \\ & \left(-r_{11}\rho^i\cos\varphi^i\sin\theta^i + r_{12}\rho^i\cos\varphi^i\cos\theta^i\right)\eta_\theta^i + \left(-r_{11}\rho^i\sin\varphi^i\cos\theta^i - r_{12}\rho^i\sin\varphi^i\sin\theta^i + r_{13}\rho^i\cos\varphi^i\right)\eta_\varphi^i + \ldots \\ & \left(r_{11}'\cos\varphi^j\cos\theta^j + r_{12}'\cos\varphi^j\sin\theta^j + r_{13}'\sin\varphi^j\right)\eta_\rho^j + \left(-r_{11}'\rho^j\cos\varphi^j\sin\theta^j + r_{12}'\rho^j\cos\varphi^j\cos\theta^j\right)\eta_\theta^j + \ldots \\ & \left(-r_{11}'\rho^j\sin\varphi^j\cos\theta^j - r_{12}'\rho^j\sin\varphi^j\sin\theta^j + r_{13}'\rho^j\cos\varphi^j\right)\eta_\varphi^j \end{aligned} \tag{18}$$

$$\begin{aligned} f_X(\varsigma) = {} & \frac{\partial f_X(\boldsymbol{\rho},\boldsymbol{\eta}',\varsigma')}{\partial \varsigma}(\varsigma-\varsigma') = \left(r_{11}\cos\varphi^i\cos\theta^i + r_{12}\cos\varphi^i\sin\theta^i + r_{13}\sin\varphi^i\right)\varsigma_\rho^i + \ldots \\ & \left(-r_{11}\rho^i\cos\varphi^i\sin\theta^i + r_{12}\rho^i\cos\varphi^i\cos\theta^i\right)\varsigma_\theta^i + \left(-r_{11}\rho^i\sin\varphi^i\cos\theta^i - r_{12}\rho^i\sin\varphi^i\sin\theta^i + r_{13}\rho^i\cos\varphi^i\right)\varsigma_\varphi^i + \ldots \\ & \left(r_{11}'\cos\varphi^j\cos\theta^j + r_{12}'\cos\varphi^j\sin\theta^j + r_{13}'\sin\varphi^j\right)\varsigma_\rho^j + \left(-r_{11}'\rho^j\cos\varphi^j\sin\theta^j + r_{12}'\rho^j\cos\varphi^j\cos\theta^j\right)\varsigma_\theta^j + \ldots \\ & \left(-r_{11}'\rho^j\sin\varphi^j\cos\theta^j - r_{12}'\rho^j\sin\varphi^j\sin\theta^j + r_{13}'\rho^j\cos\varphi^j\right)\varsigma_\varphi^j \end{aligned} \tag{19}$$

$f_X(\varsigma)$ obey gaussian distribution because of the plus character,and $E\left[f_Y(\varsigma)\right]=0$,

$$\begin{aligned} \sigma^2\left(f_X(\varsigma)\right) = {} & (r_{11}\cos\varphi^i\cos\theta^i)^2(\varsigma_\rho^i)^2 + (r_{12}\cos\varphi^i\sin\theta^i)^2(\varsigma_\rho^i)^2 + (r_{13}\sin\varphi^i)^2(\varsigma_\rho^i)^2 + (r_{11}\rho^i\cos\varphi^i\sin\theta^i)^2(\varsigma_\theta^i)^2 + \ldots \\ & (r_{12}\rho^i\cos\varphi^i\cos\theta^i)^2(\varsigma_\theta^i)^2 + (r_{11}\rho^i\sin\varphi^i\cos\theta^i)^2(\varsigma_\varphi^i)^2 + (r_{12}\rho^i\sin\varphi^i\sin\theta^i)^2(\varsigma_\varphi^i)^2 + (r_{13}\rho^i\cos\varphi^i)^2(\varsigma_\varphi^i)^2 + \ldots \\ & (r_{11}'\cos\varphi^j\cos\theta^j)^2(\varsigma_\rho^j)^2 + (r_{12}'\cos\varphi^j\sin\theta^j)^2(\varsigma_\rho^j)^2 + (r_{13}'\sin\varphi^j)^2(\varsigma_\rho^j)^2 + (r_{11}'\rho^j\cos\varphi^j\sin\theta^j)^2()\varsigma_\theta^{j2} + \ldots \\ & (r_{12}'\rho^j\cos\varphi^j\cos\theta^j)^2(\varsigma_\theta^j)^2 + (r_{11}'\rho^j\sin\varphi^j\cos\theta^j)^2(\varsigma_\varphi^j)^2 + (r_{12}'\rho^j\sin\varphi^j\sin\theta^j)^2(\varsigma_\varphi^j)^2 + (r_{13}'\rho^j\cos\varphi^j)^2(\varsigma_\varphi^j)^2 \end{aligned} \tag{20}$$

If the target is real, the target position deviation $\Delta X(i,j)$ in X axis obey Gaussian distribution with mean zero, if the target is false, the position deviation obey Gaussian distribution with mean nonzero.

The test statistic can be written as follows:

$$T = \sum_{k=1}^{N}\left(\frac{\left(\Delta X_k(i,j)\right)^2}{\sigma^2\left(f_X(\varsigma)\right)}\right) \tag{21}$$

If the target is real, test statistic T obey center Chi-square distribution with N degrees of freedom, if target is false, T obey non-center Chi-square distribution.

We choose the threshold $\eta = \chi_\alpha^2(N)$, which α is the significance level. In this situation, identifying phantom track can be regarded as the following hypothesis test:

Hypothesis H_0: If $T > \eta$, then the track is false;

Hypothesis H_1: If $T \le \eta$, then the track is true.

4 Simulation Result and Analysis

Without loss of generality, assuming that the radar network consists of three radars, and geographical coordinates of the radars are $(L_1 = 120^\circ 20', B_1 = 36^\circ 0', H_1 = 50m)$, $(L_2 = 120^\circ 5', B_2 = 36^\circ 20', H_2 = 100m)$ and $(L_3 = 119^\circ 50', B_3 = 35^\circ 40', H_3 = 200m)$, the radar sampling periods are five seconds, radar range measurement errors are 100 meters and angle measurement errors are 0.15 degrees. The starting point of false target in the radar 1 polar coordinate system is $(150000m, 30^\circ, 2^\circ)$, the finishing point is $(300000m, 180^\circ, 2^\circ)$, and the velocity of false target is 350 meters per second. Range bias error is 100 meters, angle bias error is 0.05 degrees. 31 samples are derived. the significance level is 0.05. 500 Monte-Carlo simulations have been conducted.

Without changing other parameters, the correct-discriminating probability curve along with the change of range and angle measurement error can be seen in Fig.2. Compared with conventional method in Ref.[12], the proposed method provides higher correct-discriminating probability.

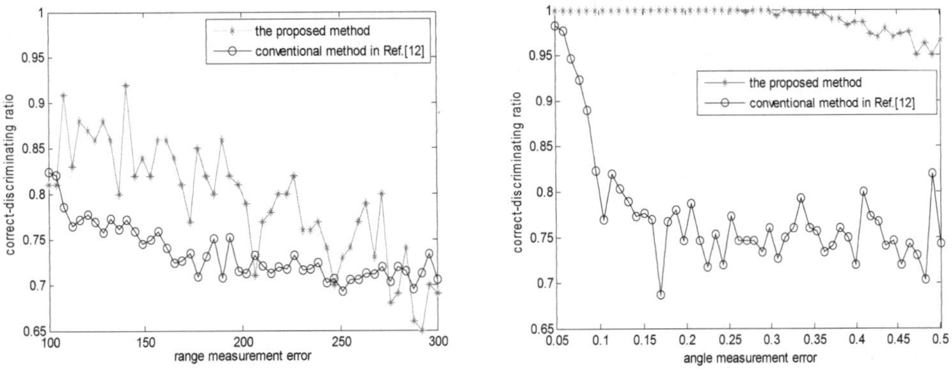

Fig.2 The correct-discriminating probability curve along with the change of measurement error

The correct-discriminating probability curve along with the change of range and angle bias error can be seen in Figure.3. From Fig.3, it can be seen that with the increase of bias error, correct-discriminating probability also increase.

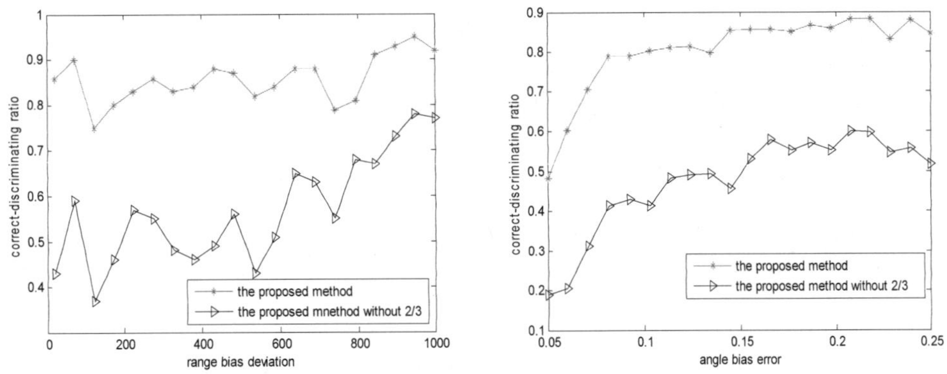

Fig.3 The correct-discriminating probability curve along with the change of bias error

5 Practical Test Results

In the test, we netted three Two-coordinate radars, and 40 samples are derived. The phantom track in radar information fusion center can be seen in Figure.4.

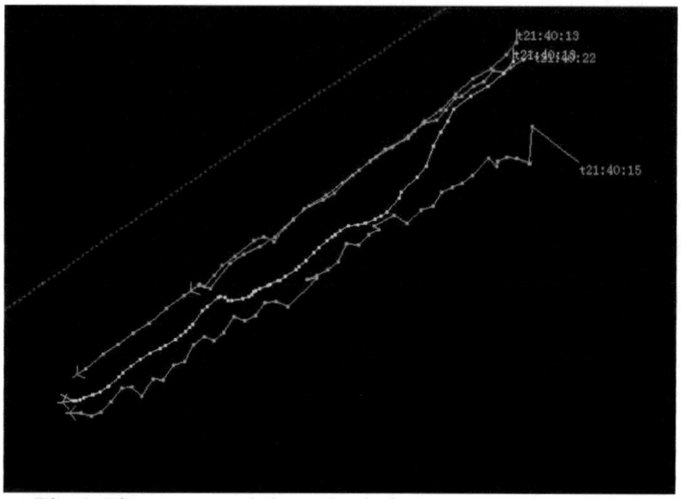

Fig.4 Phantom track in radar information fusion center

From Fig.4, the three green tracks are radar report tracks, the yellow track in the middle is fusion track. Because radar reporting track has already be filtered, we modify the radar measurement error at 1/2, the discriminating effect can be seen in Fig.5.

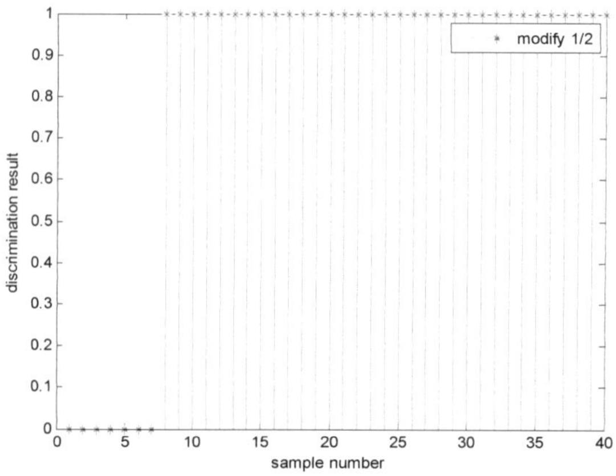

Fig.5 Discriminating effect of measurement data with modify at ½

As could be seen from Fig.5, if the effect modifies at 1/2, the proposed method can discriminate the phantom track when sample number exceed 8, so the proposed method is efficient to discriminating phantom track.

5 Conclusions

In this paper, A method for phantom track discrimination based on bias error produced by radar position error is proposed. Relations between phantom track correct-discriminating probability and radar measurement error or bias error analyzed, discrimination result of practical test proved the validity of the method, and the following conclusions observed:

(1) The correct-discriminating probability diminished with the increase of radar measurement error. When the range measurement error is 200 meters, the correct-discriminating probability of the proposed method can still be able to maintain at 0.8.

(2) The correct-discriminating probability exaggerated with the increase of root deviation which is more sensitive to angle deviation than to range deviation.

In actual combat circumstance, in order to discriminate phantom track deception released by the enemy, we should constantly changing our radar positions so that the enemy cannot precisely locate them.

References:
[1] Lee Y J, Park J R, Shin W H, et al. A study on jamming performance evaluation of noise and deception jammer against SAR satellite. *IEEE Synthetic Aperture Radar (APSAR) 2011 3rd International Asia-Pacific Conference on*. 2011; 1-3.
[2] PACHTER M, CHANDLER P R, PURVIS K B, et al. Multiple radar phantom tracks from cooperating vehicles using range-delay deception // *Proceedings of the 4th International Conference on Cooperative Control and Optimization, Destin: FL*, 2003: 1-14.
[3] B. Rao, X. S. Wang, M. Dan, et al. Ballistic Missile Tracking Algorithm with Anti-Jamming Capability in Spherical Coordinate. *Acta Electronica Sinica 2009*; 37(9): 1944-1949.
[4] F. Y. Li, J. Xu, X. D. Zhang. Pulse jamming suppression for airborne radar based on joint time-frequency analysis. *IET Radar Conference 2013, IET International*. 2013; 1-4.
[5] B. H. Wei, C. Meng, S. Y. Fan, et al. Anti-jamming Capability Evaluation of Radar Based on Fuzzy Synthetic Using Variable Weight. *Modern Radar 2010;* 32(9): 15-18.
[6] M. Pachter, P. R. Chandler, R. A. Larson, and K. B. Purvis, Concepts for generating coherent radar phantom tracks using cooperating vehicles, *in Proc. AIAA Conf. Guidance, Navigation, Control, Providence, RI, Aug, 2004, paper no*. AIAA-2004-5334.
[7] K. B. Purvis, P. R. Chandler, and M. Pachter, Feasible flight paths for cooperative gene probability of a phantom radar, *AIAA J. Guidance, Control, Dynamics*, vol. 29, no. 3, pp. 653-661, May-Jun. 2006.
[8] Purvis K B, Astrom K J, Khammash M. Estimation and Optimal Configure probability for Localization Using Cooperative UAVs, *Control Systems Technology, IEEE Transactions on Control Systems Technology*, 2008, 16(5): 947-958.
[9] ZHOU Xuli. Multiple-target track deception jamming against surveillance and warning radar, *Technology of Electronic Information Warfare*, 2007, 22(6): 43-45.
[10] CHEN Zaixing, YU Guowen, YANG Qing. Study of track deception of mono-pulse three-coordinate radar, *Journal of Air Force Radar Academy*, 2011, 25(4): 247-250.
[11] Li Fei，Zhou Zhong-liang，Gou Xin-yu,et al. Technology for penetrating radar net based on multiple combat air vehicles cooper probability track deception, *System Engineering and Electronics*. 2013，35(11):2309:2313.
[12] Sun Dian-xing, Wang Guo-hong, Zhang Xiang-yu. Algorithm of Discriminating Phantom Track for Radar Network Based on Multivariate Statistical Analysis Theory, *Acta Electronic Sinica*, 2014, 42(9): 1680-1685.
[13] WU JianPing, WANG GuoHong, SUN DianXing, et. Phantom track discrimination based in inspection of innovation in distributed radar network, *System Engineering and Electronics*. 2015, 37(1):67:72.
[14] WU JianPing, WANG GuoHong, SUN DianXing, et. A Hypothesis Testing Method for Phantom Track Discrimination in Centralized Radar Network, *Journal of XI'AN Jiao Tong University*. 2015. 49(2):80:85.
[15] He You, Xiu Jianjuan,Zhang Jingwei,et al. *Radar Data Processing With Applications*. Beijing: Publishing House of Electronics Industry, 2006.

Multi-field coupling analysis for solid rocket motor nozzle

LI yi, WU jianxin*
College of Mechanical Engineering
Inner Mongolia University of Technology
Aimin road, Xincheng District, Hohhot
China
wujx@imut.edu.cn

Abstract: Nozzle, whose performance directly determines the service life of the engine, is one of the most important part of the solid propellant rocket. The paper takes the numerical analysis of each field coupling process and analyzes the fluid-solid-thermal coupling field of the thrust vectoring nozzle of solid rocket motor. Therefore, researching on thermal effect of nozzle work and analyzing fluid-solid-thermal coupling field has become the guarantee elements which must be considered. The results show that the dynamic performance of solid rocket motor nozzle is fine. Meanwhile, the structure is reasonable. These are the basis of the development of motor axis swing TVC system in mechanical solid rocket.

Key words: Solid propellant rocket engine; Nozzle; Numeric calculation; Coupled field analysis; Flow - solid-thermal coupling; flow - solid- thermal coupling field

1 Introduction

Nozzle can provide controlled axial and lateral flight thrust for the solid rocket motor by gas to ensure aircraft's stability and run along the provision ballistic or take the maneuvering flight. Its structure is shown in Figure 1.

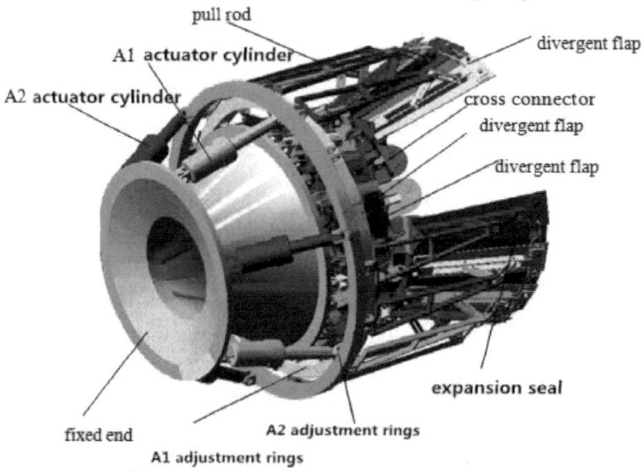

Fig.1 solid rocket engine nozzle device

On the basis of the designing parameters of solid rocket motor of the Sixth Academy of Aerospace Science belonging to China aerospace science and industry group, analyzing the transfer numerical of flow-solid, the varying coupling numerical of flow-solid and the coupling numerical of flow-solid thermal are discussed, and the data is the basis of the finite element analysis. According to the given boundary conditions, the fluid-solid-thermal coupling field of the thrust vectoring nozzle of solid rocket motor is analyzed.

2 Flow - solid- thermal coupling numerical analysis method

2.1 Thermal transfer between burning gas flow and nozzle components

The coupled thermal transfer between burning gas flow and nozzle components includes three parts: thermal conduction inside the nozzle, thermal flow and the convective thermal transfer. The numerical analysis of the loosely coupled thermal transfer between burning gas flow and nozzle components is as follows: (1) In controlling initial conditions, work out the pressure field distribution and velocity field distribution of the gas flow at time T. (2) Work out the temperature of the interface between burning gas flow field and components. (3)

Under the condition of gas flow's field and components' field at time T, calculate the values at time T+Δt. Repeat these steps and the development process of the coupled thermal transfer between burning gas flow and nozzle components could be worked out.

2.2 Strain coupling of fluidity pressure and solid components

The control of fluid diffusion equation is given as:

$$\frac{\partial \phi}{\partial t} = -u_j \frac{\partial \phi}{\partial x_j} + \frac{\partial}{\partial x_i}\left(v \frac{\partial \phi}{\partial x_i} \right) - Q - \phi \frac{\partial u_j}{\partial x_j} = R(\phi) \qquad (1)$$

The pressure of gas flow field equation is described as:

$$\Delta t^2 \theta_1 \theta_2 H \Delta \tilde{p} = \Delta t \left[G\tilde{u}^n + \theta_1 G \Delta \tilde{u}' - \Delta t \theta_1 H \tilde{p}^n - f_p \right] \qquad (2)$$

$$H = \int_\Omega (\nabla N)^T \nabla N d\Omega, \qquad (3)$$

$$G = \int_\Omega (\nabla N)^T \nabla N d\Omega, \qquad (4)$$

$$f_p = \Delta t \int_\tau N^T n^T \left[\tilde{u}^n + \theta_1 \left(\Delta \tilde{u}_1 - \Delta t \nabla p^{n+\theta_2} \right) \right] d\tau, 0 \le \theta_1, \theta_2 \le 1 \qquad (5)$$

Pressure correction equation of the speed is expressed as:

$$\Delta u = \Delta \tilde{u} - M^{-1} \Delta t \left[G^T \left(\tilde{p}^n + \theta_2 \Delta \tilde{p} \right) + \frac{\Delta t P \tilde{p}^n}{2} \right], \qquad (6)$$

$$P = \int_n (\nabla u N)^T (\nabla N) d\Omega \cdot \qquad (7)$$

Iterative computations are made based on the above equations. In addition, because the component is not flat, node method should be used in the calculations.

The processes are as followed: (1) In controlling initial conditions, work out the pressure field distribution and velocity field distribution of the gas flow at time T. (2) Load the Gas impact as equivalent surface force into the contact surface which is in contact with the components, and the displacement of the solid can be computed by its rigidity matrix, all node's loading matrix and the pressure correction equation of the speed.(3) The interface between the two fields can be built on the basis of the modular construction displacement, and its cells should be divided again. (4) Calculate the values at time T+ Δ t.

Repeat these steps and the development process of the strain coupling of fluidity pressure and the solid components could be worked out.

2.3 Numerical analysis of the fluid-solid-thermal coupling field

On the basis of the two numerical calculations above, fluid-solid-thermal coupling numeric analysis can be made. The steps are described as followed.

Step 1: in controlling initial conditions, work out the pressure field distribution and velocity field distribution of the gas flow at time T.

Step 2: analyze the temperature field of burning gas flow and nozzle components.

Step 3: the equivalent surface force load and variable temperature load are superimposed on the total load matrix. Then linear equations composed by the structural stiffness matrix, the total load array and displacement boundary condition can be calculated, and result of structural displacement can be gotten. In addition, the strain of nodes and stress matrix can be arrived at.

Step 4: based on the displacement of solid structure, rebuild the interface among the three friends. And its cells should be divided again.

Step 5: Under the new conditions, calculate the values at time T+ Δ t.

Repeat these steps and the development process of flow - solid- thermal coupling could be worked out.

3 Finite element analysis of the flow - solid- thermal coupling field

According to the design requirements, proper materials are chosen for the nozzle components. Nozzle flow field can be seen as the turbulence model. And ideal gases are assumed inside the nozzle flow field. In given the initial conditions and boundary conditions of the flow field, solid mechanics and the temperature field, finite element analysis of the coupling field can be made. The process is shown in Fig.3.

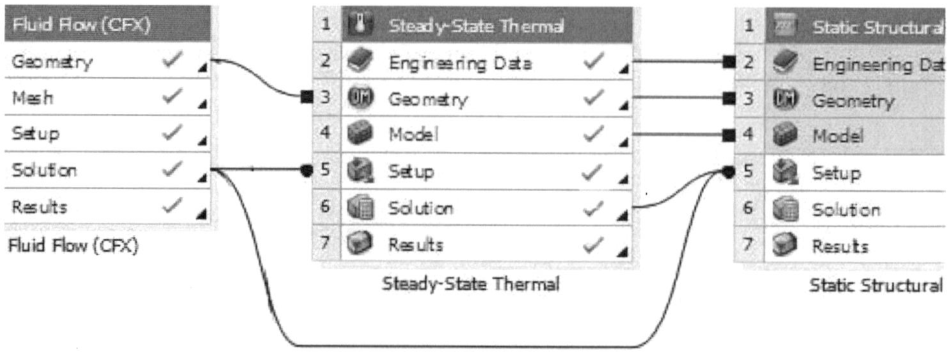

Fig.3 fluid thermal structure coupling field in the analysis process

3.1 Finite element calculation results and analysis of flow field

Analyze the flow field of simplified components, and the results show that as the fuel burns, gas flow, at the exit, present a tendency of the turbulent fluid flow. This shows that only in the flow field, the stress-strain change of the key components is not very obvious.

3.2 Finite element calculation results and analysis of flow-thermal field

Recognize the temperature load of the flow field as the boundary conditions, and then the coupling analysis can be made. Fig. 2 shows the nozzle temperature change present decreasing trend, and the temperature of the entrance is the highest. The highest temperature is 1526.9℃, and the lowest temperature is 227℃. As seen in Fig.4, the maximum deformation value of pull rod is 0.056mm, which is satisfied for material requirements. Similarly known, the maximum deformation value of the divergent flap is 0.061mm, and the maximum deformation value of convergent flap is 0.074mm. In a word, the material of divergent flap and convergent flap is satisfied for requirement.

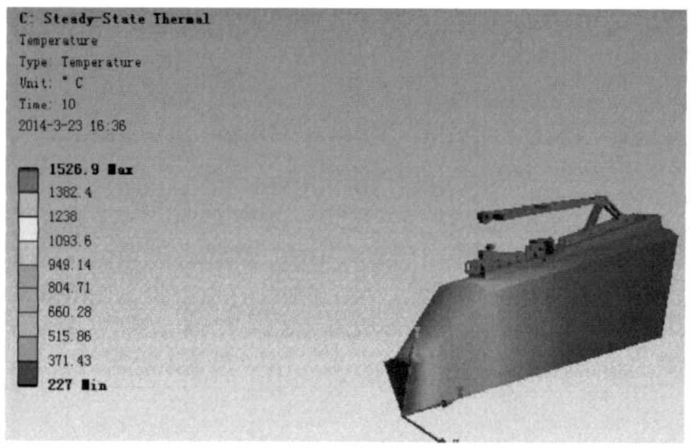

Fig.2 the flow thermal coupling field temperature change chart

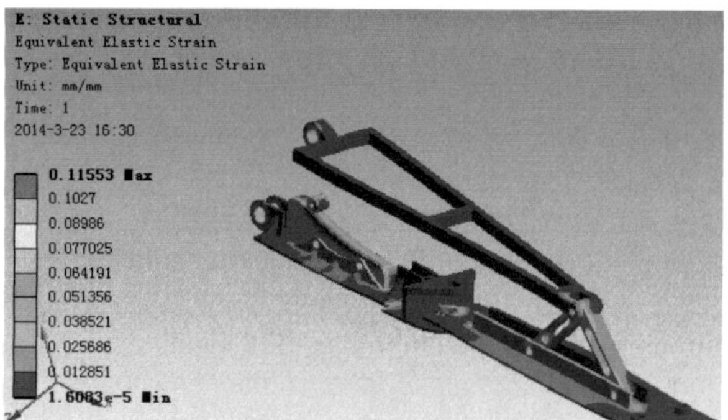

Fig. 4 the key component of strain and fluid solid coupling

3.3 Finite element calculation results and analysis of flow - solid- thermal coupling field

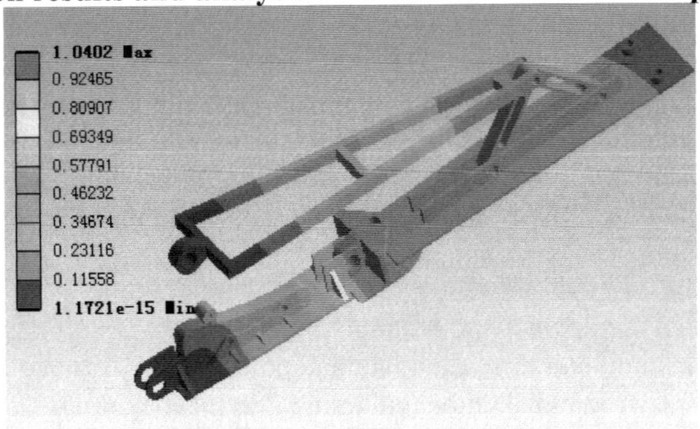

Fig.5 the strain diagram of coupling field

Based on the analysis of the coupling field in front, the finite element calculation can be made.

Fig. 5 shows maximum deformation value of the convergent flap, divergent flap and cross connector are 0.46mm, 1.0402mm, and 0.8mm. According to Fig.6, it can be concluded that the maximum stress of the convergent flap is 213.156Mpa, and meets the Requirement. Similarly, the divergent flap also has a larger safety redundancy.

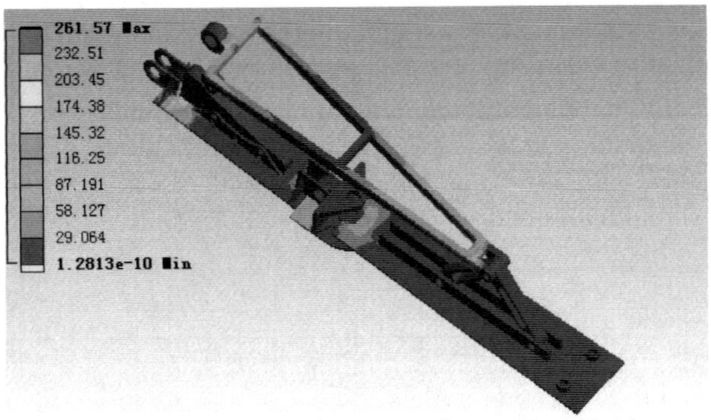

Fig.6 The stress diagram of coupling field

4 Conclusion

Based on the designing parameters of solid rocket motor of the Sixth Academy of Aerospace Science belonging to China aerospace science and industry group, the transfer numerical analysis of flow-solid, the varying coupling numerical analysis of flow-solid and the coupling numerical analysis of flow-solid thermal are discussed, and the fluid-solid-thermal coupling field of the thrust vectoring nozzle of solid rocket motor is analyzed. The results show that the dynamic performance of solid rocket motor nozzle is fine, meanwhile, the structure is reasonable.

Acknowledgment

This work was sponsored by Natural Science Foundation of Inner Mongolia (2013MS0712).

References

[1] R.A. Fiedler. Simulations of slumping propellant and flexing inhibitors in solid rocket motors [J]. Urbana-Champaign, University of Illinois Urbana, AIAA-2002-4341.

[2] A.Namazifard. Coupled Flunt-Structure 3-D Solid Rocket Motor Simulations[J]. University of Illinois Urbana, AIAA-2001-3954.

[3] Biran A, Breiner M. MATLAB 5 for engineers[M]. Boston: Addison-Wesley Publishing Company,1995: 5.

[4] Rosenfed M. An implicit solution method of multi-zone domains. AIAA, 94~0067.

[5] John Michopoulos, Charbel Farhat,Elias Houstis. Dynamic-Data-Driven Real-Time Computational

Mechanics Environment[J]. ICCS 2004, LNCS 3038,2004,693-700.
[6] J.Michopoulos, C. Farhat, E. Houstis, etc. Dynamic Data Driven Methodologies for multiphysics system modeling and simulation. ICCS 2005, LNCS3515, 2005,616-623.
[7] J.Michopoulos, P. Tsompanopoulou, E. Houstis. On a data-driven environment for multiphysics applications. Future generation computer systems. 2005(21):953-968.
[8] A.V. Smirnov. Multi-physics modeling environment for continuum and discretedynamics. 2004,1-12.[41] T.Belytschko, R.Mullen.Stability of explicit-implicit mesh partitions in time integration. Int. J. Number. Methods Emgrg.,1978(12):1575-1586.
[9] O.Sigmund. Design of multiphysics actuators using topology optimization-Part One-material structures. Computer methods in applied mechanics and engineering, 2001(190) :6577-6604.
[10] O.Sigmund. Design of multiphysics actuators using topology optimization-Pa II :two-material structures. Computer methods in applied mechanics and engineering,2001(190):6605-6627.
[11] Salvetat B. Analysis of Gas Flow in Three Dimensional Solid Rocket Propellant Grains[J]. AIAA 84~1357.

Femtosecond laser machining SiC surface roughness control model research

Li Xiaoyu, Sun Huilai, Yu Xinxin
School of Mechanical Engineering
TianJin Polytechnic University
No.399, West of Binshui road, Xiqing district, Tianjin City, CHINA
owokan@sina.com

Abstract: Research on mechanism of femtosecond laser interaction with SiC is fully considering the influence of many factors within the scope of the laser pulse width and the decline of free electron density factor as considering the femtosecond laser beam in the process of the processed material transmission losses. They caused by processed material surface reflection; and due to the collision after laser ionization photoelectron and lattice energy loss as "reflection and scattering". So, this paper is proposing the existing mechanism for correction of the improvement.

Key words: femtosecond laser; SiC; silicon carbide; roughness; pulse train

1. Introduction

Femtosecond laser processing is a kind of important micro-nano processing field processing method, its processing and application scope is widespread, from the soft metal, such as the hardened steel, stainless steel, high speed steel, cemented carbide and other hard processing material, to the semiconductor, glass, ceramics and other hard brittle non-metallic materials[1]. Femtosecond processing important features is the small scale by machining, machining accuracy high, almost no heat affected zone, etc[2~3]. With modern industrial technology and high performance products for parts precision and surface, integrity of the demand is higher and higher, the role of micro-nano processing is becoming increasingly important[4~6]. Its defense, aerospace, nuclear power and other high-tech areas also has important influence, therefore it attaches great importance to by each industrial developed countries[7].

Represented by silicon carbide (SiC) of wide band gap, semiconductor has low thermal expansion coefficient, high temperature resistance, excellent chemical stability, good electrical insulation, low and stable ultrasonic delay performance, best ultraviolet spectrum performance as well as visible light and near infrared spectrum properties and mechanical properties which is higher than common glass[8].

2. Femtosecond laser and action mechanism of SiC materials

Through the study of femtosecond laser, machining SiC is preliminary mastered by its interaction mechanism. Through Experiment with 800 nm Ti: Sapphire femtosecond laser, spot diameter 1 um, pulse width 120 fs, laser power density of 15.28 J/cm2, through the analysis of test results, we found that in the process of t = 50 fs to 80 fs, corrosion pit width and depth are growing rapidly, but the width speed increase is significantly higher than the depth increase, as the situation of corrosion pits on the type of flat structure, to improve surface roughness of femtosecond laser machining.

In the field of traditional processing, reducing the cutting parameter can effectively improve the machining precision and quality materials. Scholars at home and abroad, through the experiment in the field of femtosecond laser machining got the same result. In the process of femtosecond, as long as the pulse energy is not lower than the threshold needed for processing, it can achieve the purpose of processing, then divide the single pulse energy for n after pulse sequence, then equal to the decrease in the process of machining cutting parameter, thus it achieves the goal of improving the material processing precision and quality. Now through the experiment on the statement (Fig 2), however, the surface roughness research lack theoretical support in material processing [9-10].

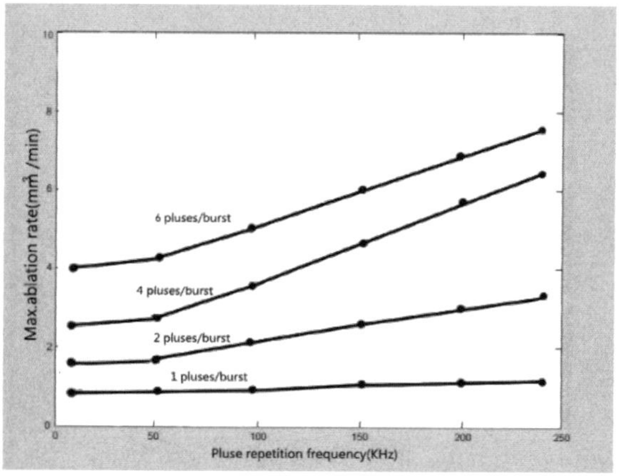

Fig 1. Processed materials (SiC) the relationship between the removal rate of removal rate and pulse

Institute of German Lumera Laser GmbH Ralf Knappe et al [11] to nickel base of blade in the process of punching processing experiments, improve roughn- ess . table seconds need appropriate strategy which can achieve a good surface roughness. Blade thickness is 1 mm, pulse sequence number is 10, pulse repetition frequency is 1 MHz, machining the hole with a diameter of 400 um is into 60 degrees.

3. Theoretical model improvement research

3.1 Assumptions

In the process of the interaction between femtosecond laser machining SiC, we fully considered the femtosecond laser and substrate material in the process of ionization. Due to the influence of many factors, the factor $\frac{n_e(t,r,z)}{\tau}$ which causes ionization which reduces numbers of electrons, Femtosecond laser beam is in the process of the processed material transmission,.The yield loss caused by processing material surface reflection $\int_0 R(t,r,0)dz$, and scattering occurs, result in the energy loss of $\int_0 \frac{n_e(t,r,z)}{\tau}dz$, thus to improve the existing mechanism.

3.2Femtosecond laser processing control model of surface roughness of SiC materials

Femtosecond laser beam which is the role of SiC materials is free electron ionization process of substrate material. When the free electron density reaches, the critical threshold of substrate material remove begin to be processed, substrate material and laser beam produced by the plasma oscillation frequency equals to the femtosecond laser frequency, using Flux - Dubling model and Fokker-Plank formula to calculate the free electrons generated[12~14]:

$$\frac{\partial n_e(t,r,z)}{\partial t} = \alpha_i \times I\ (t,r,z) \times n_e(t,r,z) + \delta_N \times I^N(t,r,z) - \frac{n_e(t,r,z)}{\tau} \qquad (1)$$

Laser in the processed substrate material internal distribution function is calculated as follows:

$$I(t,r,z)=\frac{2F}{\sqrt{\pi/\ln 2t_p}}\times[1-R(t,r,z)]\times\exp\left\{-\frac{r^2}{r_0^2}-4\times\ln 2(\frac{t}{t_p})^2-\int_0^z[\alpha(t,r,z)+R(t,r,0)+\frac{n_e(t,r,z)}{\tau}]dz\right\} \quad (2)$$

According to the improved Gauss beam laser, the intensity distribution inside the characteristic analysis of processed materials, the distribution of femtosecond laser intensity is calculated, and the processed material dielectric time space function expressions is calculated, and then the optical properties derived A and B is calculated, which is calculated as[15]:

$$R(t,r,z)=\frac{[f_1(t,r,z)-1]^2+f_2^{\ 2}(t,r,z)}{[f_1(t,r,z)+1]^2+f_2^{\ 2}(t,r,z)} \quad (3)$$

$$\alpha(t,r,z)=[\alpha_i\times ne(t,r,z)+\delta_N I(t,r,z)^{N-1}]\times[\varepsilon(t,r,z)+U_I] \quad (4)$$

By calculating the largest minimum electrons and photons - ion collision ionization parameters, we calculated Clouomb logic value, calculated as[16]:

$$\ln\Lambda = \frac{1}{2}\times\ln\left[1+\left(\frac{b_{\max}}{b_{\min}}\right)^2\right] \quad (5)$$

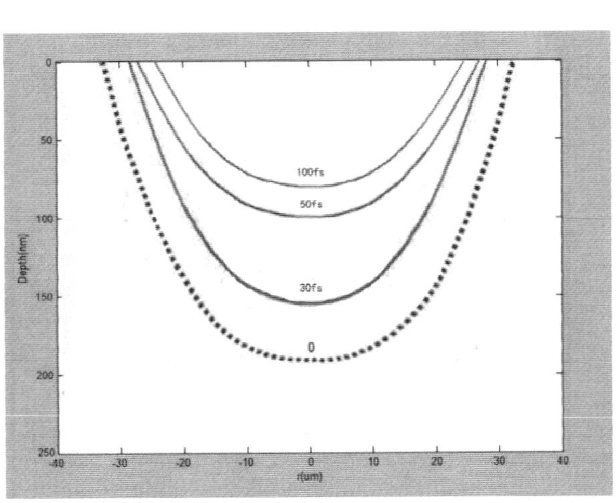

Fig 2. Pulse ablation under 50fs pit shape (laser power 5 J/cm2)

At the same time, the initial reaction stage of the laser and electron electronic kinetic energy are relatively low. Free electrons in the process of heating time, the electronic - photon collision impact are obvious. Therefore, in the process of calculation of the free electron heating time, we must consider the influence of electrons and photons and electronic - ion collision. Finally, we got the free electrons in the process of femtosecond laser when machining heating time numerical model[17], mathematical expressions as shown in the formula (6).

$$\tau_e(t,r,z)=\frac{3\sqrt{m_e}[K_B T_e(t,r,z)]^{\frac{3}{2}}}{2\sqrt{2}(Z^*)^2 n_e(t,r,z)e^4\ln\Lambda}\times\left\{1+\exp\left[\frac{-\mu(n_e,T_e)}{K_B T_e(t,r,z)}\right]\right\}F_{\frac{1}{2}} \quad (6)$$

Free electron temperature is obtained by the above numerical model, and the free electron special heat (mathematical expressions as shown in formula 6,7), free electronic absorption, chemical potential (mathematical expressions as shown in formula 8[18]), Fermi energy (mathematical expressions as shown in formula 9),a single electron average kinetic energy can be calculated by using the Fermi Dirac distribution - electronic average, and then can get out of the electronic - photon heat numerical model[19] as:

$$C_e(T,n_e)\frac{\partial T(t,r,z)}{\partial t} = \nabla[k_e(T)\nabla T] - G(T - T_l) + S(t,r,z) - H_e(t,r,z) \quad (7)$$

$$\mu(T,n_e) = \varepsilon_F(n_e)\times\left\{1-\frac{\pi^2}{12}\times\left[\frac{\kappa_B T(t,r,z)}{\varepsilon_F(n_e)}\right]^2+\frac{\pi^2}{80}\times\left[\frac{\kappa_B T(t,r,z)}{\varepsilon_F(n_e)}\right]^4\right\} \quad (8)$$

$$\varepsilon_F(n_e) = \left(\frac{(hc)^2}{8m_e c^2}\right)\left(\frac{3}{\pi}\right)^{2/3}(n_e(t,r,z))^{2/3} \quad (9)$$

For femtosecond laser, machining depth has more mature theoretical basis, but it is not unified $d = (1/\alpha)\times\ln(F_0/F_{th})$ or $h = h_0 \times \ln(F_0/F_{th})$ .For machining width , as well as the corrosion pit at the bottom of the shapes which have no unified theory model. This paper is based on the interaction between femtosecond laser and wide band-gap materials mechanism of temperature distribution and improved gaussian distribution characteristics. Corrosion pit structure characteristics of the forecast model set up; Femtosecond laser and material in the process of interaction mechanism, are based on the laser intensity distribution in accordance with the modified gaussian distribution (i.e., considering the lattice reflection and scattering of laser energy density), It is due to the temperature distribution T (t, r, z) function, so in the femtosecond laser pulse intensity, it is greater than or equal to threshold intensity, whose material is removed for processing.

$$\begin{cases} I_f(t,r,z) \ge n_{cr} \\ C_e(T,n_e)\frac{\partial T(t,r,z)}{\partial t} = \nabla(k_e(T)\nabla T) - G(T-T_l) + S(t,r,z) - H_e(t,r,z) \end{cases} \quad (10)$$

Simultaneous solution of formula (10), get corrosion pits structure characteristic value r_c across the width and the depth of numerical z_c direction, and the relationship between r_c and z_c expression $R_{r\ z}(\tau_e,r,z)$, expressed as follows:

$$\begin{cases} r_c = \frac{\partial T(t,r,z)}{\partial r} \\ z_c = \frac{\partial T(t,r,z)}{\partial z} \end{cases} \quad (11)$$

$$R_{r\ z}(\tau_e,r,z) = \frac{\partial T(t,r,z)}{\partial r \partial_z} \quad (12)$$

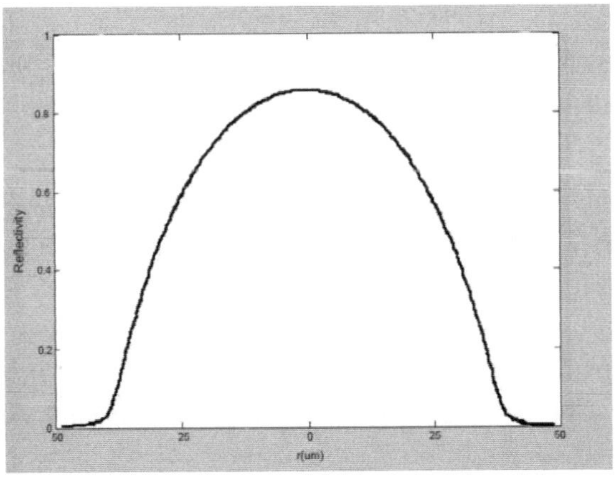

Fig 3. Relationship between reflectivity of SiC material surface and laser radius

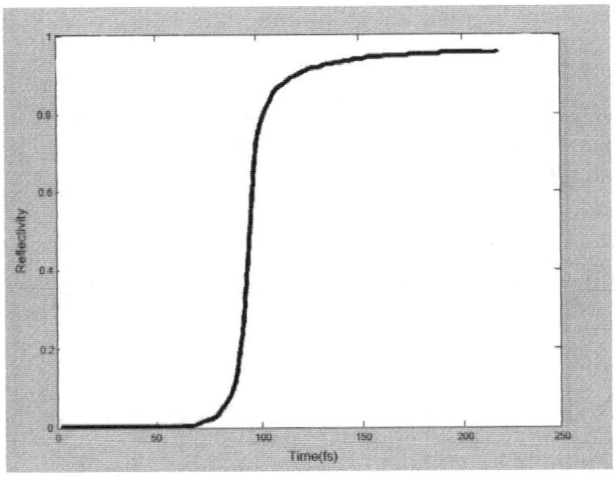

Fig 4. Relationship between reflectivity and pulse period of SiC material surface

4. Conclusion

On the basis of the previous work, we can get tight femtosecond laser single pulse processing wide band material surface roughness value and productivity of processed materials (volume/mass productivity, namely corrosion pit volume/quality and pulse duration ratio).Based on the structural characteristics and processed material ,Machining processing pulse energy threshold can be needed. Under the condition of more than threshold, we can cut a single pulse energy (energy redistribution) which is divided into pulse sequence, to reduce the surface roughness.

Meet the conditions of $z(\tau_e, r, z) = \dfrac{\partial R_{r\ z}(\tau_e, r, z)}{\partial r} \le z_0$,could be assigned single pulse/pulse sequence processing to meet the requirements of surface roughness. Femtosecond laser beam parameters, the corresponding $r(\tau_e, r, z_0) = \dfrac{\partial R_{r\ z}(\tau_e, r, z_0)}{\partial r}$, the femtosecond laser machining overlap rate O of wide band gap materials are as follows:

$$O = \frac{r(\tau_e, r, z) - r(\tau_e, r, z_0)}{r(\tau_e, r, z)} \tag{13}$$

References

[1] M. Lenzner, J. Krüger, S. Sartania, Z. Cheng, C. Spielmann, G. Mourou,W. Kautek, and F. Krausz, *Phys. Rev. Lett.* 80, 4076 _1998_.

[2] R. Le Harzic, N. Huot, E. Audouard er al.. *Comparison of heat-affected zones due to nanosecond and femtosecond laser pulses using transmission electronic microscopy[J].* Appl. Phys. Lett., 2002.,80(21): 3886~3888

[3] S. Nolte, C. Momma, H. Jacobs et al.. *Ablation of metals by ultrashort laser pulses[J].* J. Opt. Soc. Am. B: Opt. Phys., 1997, 14(10): 2716~2722

[4] R. Gattass, E. Mazur. *Femtosecond laser micromachining in transparent materials [J].* Nat. Photon., 2008, 2(4): 219~225

[5] J. X. Gong, X. Zhao, Q. Xing et al.. *Femtosecond laserinduced cell fusion [J].* Appl. Phys. Lett., 2008, 92(9): 093901

[6] S. Kawata, H. B. Sun, T. Tanaka et al.. *Finer features for functional microdevices-micromachiner can be created with higher resolution using two-photon absorption* [J]. Nature. 2001, 412(6848): 697-698

[7] CAPPELLI E，ORLANDO S，SCITI D，et al. *Ceramic surface modifications induced by pulsed laser treatment*[J]. Appl. Sur. Sci.，2000，154-155：682-688.

[8] ZOPPEL S，FARSARI M，MERZ R，et al. *Laser micromachining of 3C-Si C single crystals*[J]. Micro- electronic Engineering，2000，83：1400-1402.

[9] Mangirdas Malinauskas, Paulius Danilevi cius, and Saulius Juodkazis. *Three-dimensional micro-/nano-structuring via direct write polymerization with picosecond laser pulses.* OPTICS EXPRESS.2011,19(6):5602-5610

[10] B. Rethfeld, A. Kaiser, M. Vicanek, G. Simon, *Ultrafast dynamics of nonequilibrium electrons in metals under femtosecond laser irradiation*, Phys. Rev. B 65 (2002) 214303–214313.

[11] Ralf Knappe, Hatim Haloui, Albert Seifert. *Scaling ablation rates for picosecond laserScaling lasers using burst micromachining.* Lumera Laser, 2010.

[12] B. C. Stuart, M. D. Feit, S. Herman, A. M. Rubenchik, B. W. Shore, and M. D. Perry, *Phys. Rev. B* 53, 1749 _1996_.

[13] B. C. Stuart, M. D. Feit, A. M. Rubenchik, B. W. Shore, and M. D. Perry, *Phys. Rev. Lett.* 74, 2248 _1995_.

[14] M. D. Perry, B. C. Stuart, P. S. Banks, M. D. Feit, V. Yanovsky, and A. M.Rubenchik, J. *Appl. Phys.* 85, 6803 _1999_.

[15] L. Jiang and H. L. Tsai, J. Phys. D 37, 1492 _2004_.L. Jiang, H.L. Tsai, J. *Phys. D: Appl. Phys.* 37 (2004) 1492–1496

[16] K. Eidmann, J. Meyer-ter-Vehn, T. Schlegel, and S. Hüller, *Phys. Rev.* E62, 1202 _2000_.

4]B. Rethfeld, A. Kaiser, M. Vicanek, and G. Simon, *Phys. Rev. B* 65,214303 _2002_.

[17] Lee Y T and More R M 1984 *Phys. Fluids* 27 1273. _1984_.

[18] N. W. Ashcroft and N. D. Mermin, *Solid State Physics _Holt, Rinehart, and Winston, New York,* 1976_.

[19] L. Jiang and H. L. Tsai, ASME J. Heat Transfer 127, 1167 _2005.

AUTHOR INDEX

Imperial Consortium of Research and Development (ICRD)
P.O. Box: 116974
Dubai, United Arab Emirates

ISBN 978-1-5108-3260-2
9 781510 832602

CURRAN ASSOCIATES INC.
proceedings.com